General Relativity Revisited

The Speed Of Light

by

Patrick John Naughton

Patrick Naughton

Contents

Patrick Naughton

Introduction

The aim of this text is to present an alternative definition of General Relativity. The fundamental idea of General Relativity was derived from ideas first proposed by Albert Einstein back in 1905, when he was just 25yrs old. This has been a hugely successful theory which has substantially helped us to better understand the world at large. However, despite the immense success of Einstein's original theory, its application does appear to lead to some dilemmas and inconsistencies. Einstein continued his work right up until the last few months of his life, fully aware that there was still more that could be done to develop his ideas of General Relativity still further.

I've attempted to build on Einstein's theory and present a slightly modified alternative formulation. However, it's important to make clear that the ideas presented here are based purely on conjecture and are put forward simply to promote further discussion. These proposals are relatively simplistic and are not in any way proven, but may hopefully help to promote discussion and perhaps inspire further improvements.

Patrick Naughton
July 2024

Patrick Naughton

General Relativity Revisited

1. Einstein's General Relativity

According to Einstein, the Kinetic Energy of a moving object of stationary mass m is given by the formula;-

$$E = mc^2(Y-1) \qquad ...(1)$$

Where c is the speed of light and Y is the Lorentz factor defined as;-

$$Y = 1/(1 - V^2/c^2)^{0.5} \qquad ...(2)$$

where V is the speed of the moving object and c is the speed of light .
If this expression for energy E is expanded as a series then we get;-

$$E = mc^2(1 + 1/2V^2/c^2 + 3/8V^4/c^4 +15/48V^6/C^6 + \cdots) - mc^2 \qquad ...(3)$$

The Lorentz factor Y is significant for objects travelling at very high speeds compared to the speed of light, when the impact of General Relativity becomes more noticeable. At normal lower speeds, when V is much lower than the speed of light, then the Y factor converges to a value of unity (Y = 1) as the velocity V is reduced towards zero.
At low velocities then the values of V^4/c^4 and higher powers of V/c become insignificant. When this is the case then the traditional classical definition of Kinetic Energy agrees with Einstein's definition of Energy;-

$$KE = \frac{1}{2} mV^2 \qquad ...(4)$$

At very high velocities, as V approaches the speed of light c then the value of the Lorentz coefficient Y approaches infinity Y=>∞ This means that the Kinetic Energy of a speeding object tends towards infinity as the speed of the object approaches the speed of light V = C which Einstein proposes is the maximum speed that can ever be achieved. It's for this reason that the speed of light c is often referred to as the universal speed limit

Patrick Naughton

Some Potential Problems With Einstein Relativity

Einstein's theory can be easily derived (see appendix 1). However there appear to be some questionable aspects of this simple derivation. For example, Einstein himself tells us that mass causes the dimensions of spacetime to be curved. However, the derivation of Einstein's relativity depends wholly on the Pythagoras theorem. Unfortunately, the Pythagoras theorem is only valid on a flat surface where the internal angles of a ninety-degree triangle add up to 180° degrees.

This seems to be something of a weakness in Einstein's derivation. It's quite possible for example, to draw a triangle on a curved surface where *all three angles* are ninety degrees, so the Pythagoras theorem would no longer apply. If we were to draw lines on the surface of the Earth from the North Pole to the equator, it's possible to draw a triangle (albeit not a flat one) with three ninety-degree angles. It seems at the very least slightly questionable that Einstein's relativity relies on the existence of a flat surface which may not necessarily exist in relativistic spacetime when viewed by an observer external to the timeframe in question.

Another aspect of Einstein's relativity which seems a potential area of weakness is the formula for the Lorentz factor Y. This equation for Y mandates that the maximum relative speed a particle can reach is 'c' the speed of light. According to Einstein, the speed of light is in effect the Universal speed limit. There does seem to be something of a problem with this assertion. Think about our position in the universe and how we observe the stars. Here on the surface of the Earth, we are turning at a rate of about 360° degrees every 24 hours or thereabouts. That applies to everybody who stands on the surface of the Earth, whether they're on the equator or shivering somewhere up near the frozen arctic poles. That means that anything more than about 2.8 billion miles away is moving *relative to us* at a speed greater than the speed of light, which includes every single star other than our own sun. The nearest star to us beyond our sun, is Alpha Centauri, which is about 25 trillion miles away. In this regard **every single star, except our own sun, appears to violate Einstein's maximum speed limit**, at least from the point of view of everyone on our spinning Earth.

General Relativity Revisited

These 'observations', along with similar questions give some cause to doubt the total reliability of the current widely accepted form of Einstein's relativity. This work was carried out to see if an alternative derivation of relativity could be achieved.

2. The Orbit Of A Single Electron In An Atom

Objects that move at low speeds compared to the speed of light do not appear to demonstrate pronounced effects of relativity. In order to investigate the effects of relativity, it was thought best to consider objects that are moving at higher speeds somewhere approaching the speed of light. I decided to analyze the orbit of a single electron in an atom in order to keep the calculations as simple as possible.

Atoms of the different elements contain different numbers of electrons. In their normal state the number of protons in the nucleus of an atom equals the number of electrons orbiting in the atom. Electrons can be removed from any atom by ionisation. If a photon of light with energy exactly equal to the Kinetic Energy of an electron in an atom is absorbed by the electron, then that electron can be liberated from the atom by ionization. For the purposes of this work, it was assumed that electrons orbit in circular orbits around the nucleus of the atom and the electrons are assumed to be equivalent to small ball bearing type objects concentrated in one spherical point. It is recognised this is not a model based on reality but it is thought to be a sufficiently convenient model to determine adjustments to the fundamental definition of General Relativity.

This simplified representation of the orbit of the last remaining electron in an atom is shown in diagram 1 below. In this diagram the last remaining electron in the atom is shown with charge -e travelling at velocity V at radius L around the nucleus of the atom with P protons with charge +P x e.

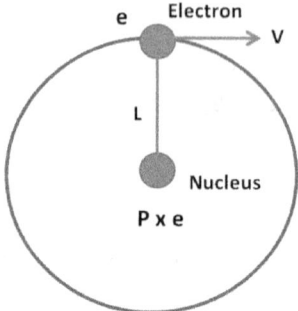

Diagram 1 – A simple model showing the last electron in an atom with P protons in the nucleus travelling in a circular orbit with radius L.

The single orbiting electron is attracted to the protons P in the nucleus of the atom. The electrostatic attraction of the electron to the protons is given by Coulomb's force F_r where;-

$$F_r = Pe^2/4\pi\varepsilon L^2 \qquad \qquad ...(5)$$

In this equation ε is the permittivity of free space and the radius of the atom with one remaining electron is L. This is a simple classical Bohr model.

The energy Kinetic Energy E of the electron in the orbit is given by the formula;

$$E = hf \qquad \qquad ...(6)$$

where h is Planck's constant and f is the frequency of the photon of light needed to liberate the electron from the atom by ionization.

The orbiting electron has a deBroglie wavelength which is given by the formula;-

$$\lambda = 2\pi LS' \qquad \qquad ...(7)$$

where S' is the reciprocal of the spin of the electron.

In 1913 Physicist Niels Bohr proposed a model for the atom which meant that electrons could only orbit atoms in certain discrete orbits. Bohr also argued strongly that the electron had to orbit in such a way as to complete standing deBroglie waves.
The spin of a free electron is nominally taken to be spin = ½ therefore S' would nominally be S' ~ 2

We know from measurements of free electrons that;

$$S'=2.0023193$$

If the value of S' was exactly 2 this would effectively mean that the electron would need to complete two orbits of the atom in order to complete one full wavelength as illustrated in the diagram below;-

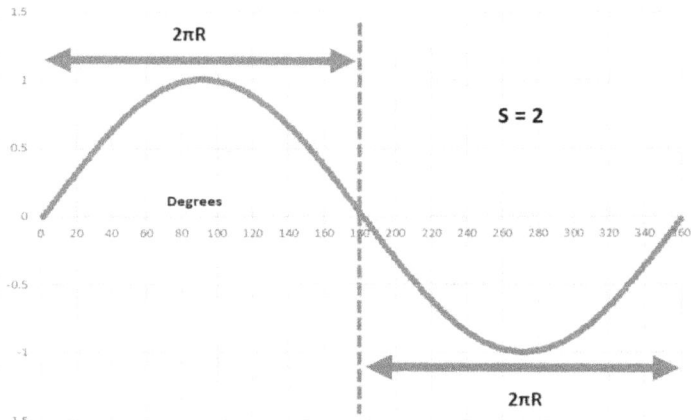

Diagram 2 – An electron with spin = ½ would need to complete two full orbits of an electron to complete one wavelength.

The velocity of a wave is related to the wavelength λ by the formula;-

$$V = f \times \lambda \qquad\qquad ...(8)$$

7

But we know that distance is reduced at speed by the Lorentz factor Y (although Einstein appears to ignore this in his calculation of Y)

So we can say

$$\lambda = 2\pi x L x S'/Y \qquad ...(9)$$

Putting together equations 6, 7, 8 and 9 we can say;-

$$E = (hxVxY)/(2\pi S'L) \qquad ...(10)$$

Rearranging gives;-

$$L = (hxVxY)/(2\pi S'E) \qquad ...(11)$$

Putting this result in equation 5 we can say;-

$$F_r = Pe^2 \pi S'^2 E^2/(\epsilon h^2 V^2 Y^2) \qquad ...(12)$$

It is widely accepted that the Centripetal Force F_C that has to operate on any object in order to make it travel in a circular orbit is given by the formula;-

$$F_C = YmV^2/L \qquad ...(13)$$

where;

m is the mass of the object travelling in a circular orbit
V is the speed of the object
L is the radius of the circular orbit
and Y is the Lorentz coefficient (although at this stage the exact definition of Y remains open.)

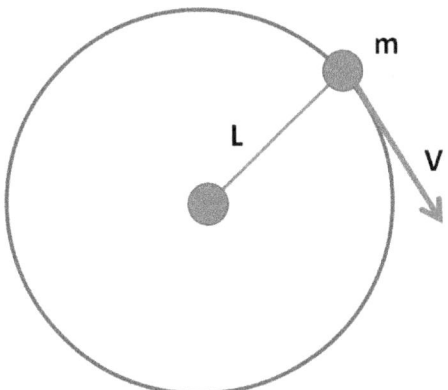

Diagram 3 – An object of mass m travelling at velocity V in a circular orbit of radius L.

At speeds much lower than the speed of light then as $Y => 1$ then equation 11 becomes equivalent to the classical formula for centripetal force $F_c => mV^2/L$

For a single electron travelling in an approximately circular orbit in an atom we can assume that the centripetal force F_c is created by the Coulomb force F_r

Therefore from equations 5 and 13 and assuming L and R are the same in a circular orbit we can say;-

$$YmV^2 = Pe^2/4\pi\varepsilon L \qquad\qquad ...(14)$$

Re-arranging equation 14 and substituting the expression for L from equation 11 we can say;-

$$V_2^3 = Pe^2 x E x S'/2h\varepsilon mY^2 \qquad\qquad ...(15)$$

I'll refer to this velocity as Naughton velocity and refer to it as velocity V_2. What is particularly important about this formula is that it doesn't in any way rely on the _definition_ of the Lorentz factor Y expressed in terms of V as used by Einstein and stated in equation 2. The values of the energy E of the last electron in each atom are well known and are listed in appendix 2.

We can easily calculate the values of the Lorentz factor Y from the formula;

$$E = mc^2(Y-1)$$

From these values of Y and the ionization energy E of the last electrons and values of protons P for each atom it is possible to calculate values for V_2 for the last electrons in each atom.

Quick Recap - What Have We Achieved?

We know the energy required to ionize the last electron out of each atom. From this energy E we can calculate the actual value of the Lorentz factor Y in each case. And thanks to equation 15 we can calculate the value V_2 the speed of the last electron in each atom.

By plotting 1/Y v $(V/C)^2$ it's therefore possible to get a slightly modified expression for Y

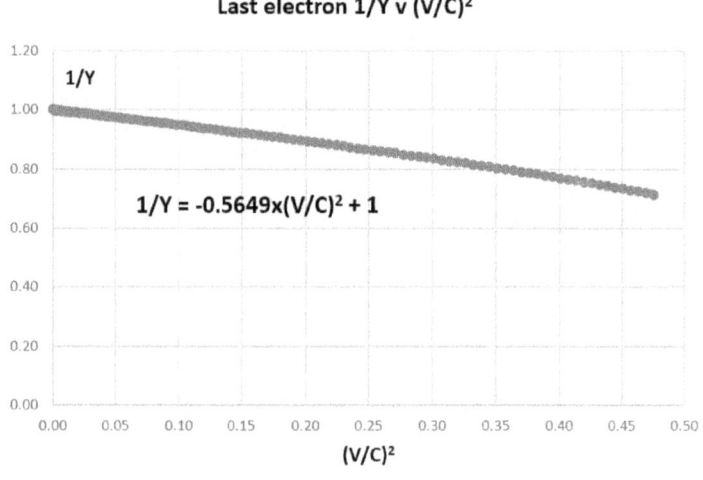

Last electron 1/Y v $(V/C)^2$

$$1/Y = -0.5649x(V/C)^2 + 1$$

Graph 1 - 1/Y v $V_2{}^2/C^2$ for the last electron in each atom

General Relativity Revisited

This simple graph gives a new modified expression for the Lorentz factor;-

$$Y = 1/(1 - 0.5649V^2/C^2) \qquad \dots(16)$$

An Inconsistency In Einstein Relativity

Having derived this alternative definition of the Lorentz factor, it is interesting to compare this to the Einstein definition to determine which solution best agrees with observations.

From equation 15 and knowing the ionisation energy E and therefore, also the value of the Lorentz coefficient for the last electron in each atom, it's possible to calculate their velocity if they were to move in a circular orbit in each atom. (Whether these electrons *actually* travel in a perfectly circular orbit is largely irrelevant, but it can be assumed they do so as the energy required to ionize them out of their host atom remains consistent.)

By plotting the graph of V/C where V is the velocity of the last electron in each atom against P the number of protons in each atom, we get the following results;-

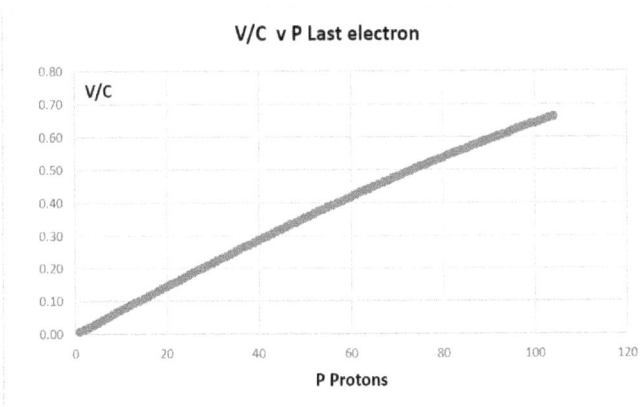

Graph 2 - V/C v P protons for the last electron in each atom.

If we look closely, we can see this graph has a slight curve, and as such we can reasonably expect the value of the velocity V (or the Lorentz coefficient Y) to reach a peak at some point when the value of P is large enough. So, let's calculate at what point the peak is reached in the case of Einstein's definition of Y and the Naughton definition of Y.

The Einstein Peak

From equation 15 we know that

$$V^3 = Pe^2xExS'/2h\epsilon mY^2 \qquad \text{...(17)}$$

Remember, this formula *does not depend on any particular definition* of Y

Let's apply the Einstein definition of Y from equation 2

$$Y = 1/(1 - V^2/c^2)^{1/2} \qquad \text{...(18)}$$

Re-arranging this formula we can say;

$$V^2 = c^2 (Y^2-1)/Y^2 \qquad \text{...(19)}$$

or

$$V^3 = c^3 (Y^2-1)^{3/2}/Y^3 \qquad \text{...(20)}$$

Putting this Einstein expression into equation 17 and re-arranging gives;-

$$P = (2h\epsilon c/S'e^2) \times (Y - 1)^{1/2} \times (Y + 1)^{3/2} \times Y^{-1} \qquad \text{...(21)}$$

Differentiating with respect to Y gives...

$$dP/dY = (2h\epsilon c/S'e^2)x[(Y+1)^{1/2}x(Y^2-Y+1)x(Y+1)^{1/2}]/[(Y-1)^{1/2}xY^2] \quad \text{...(22)}$$

There is never a point where (Y^2-Y+1) or any other factor in this equation causes dP/dY to reach zero (a maximum turning point) when Y >= 1 So

12

the Einstein definition of Y can never yield a maximum value of P in contradiction to what is indicated in Graph 2.

The Naughton Peak

From equation 15 we know that

$$V^3 = Pe^2xExS'/2h\epsilon mY^2 \qquad ...(23)$$

Using the definition for energy $E = mc^2(Y-1)$ from equation 1 we get;-

$$V^3 = Pe^2xc^2(Y-1)xS'/2h\epsilon Y^2 \qquad ...(24)$$

If we say $A = 2h\epsilon/(e^2xc^2xS')$ and substitute the Naughton definition for Y from equation 16 then we get;-

$$AV^3 = P \times aV^2/c^2 \times [1 - aV^2/c^2] \qquad ...(25)$$

where a = 0.5649

Differentiating both sides with respect to V gives;-

$$6AV^2 \times dV/dP = \{2aV/c^2 - 4a^2V^3/c^4\} + P\times[2a/c^2 - 12a^2V^2/c^4] \times (dV/dP) \quad ...(26)$$

when $dV/dP = 0$ then...

$$2aV/c^2 = 4aV^3/c^4 \qquad ...(27)$$

so a turning point occurs when

$$V = 0.9408c \text{ at which point } P \sim 168 \qquad ...(28)$$

The important point is that a turning point is predicted by the Naughton version of the Lorentz coefficient Y which does not occur in the case of the Einstein definition.

This indicates there is a maximum turning point in the V/c v P relationship resulting from the Naughton definition of Y which is unexpectedly absent in the case of the Einstein definition of Y.

As one might expect (since equation 15 was in part used to derive the Naughton-Lorentz definition) this formula accurately predicts the atomic numbers of the elements, when values for the ionisation energies of the last electrons are fed in as long as the Naughton-Lorentz definition of Y is used. For example, the ionisation energy of the last electron Rutherfordium with atomic number 104 (the element with the highest atomic number for which energy data was used in this work) is 177.148eV

When this value is fed into equation 15 and the Naughton-Lorentz definition of Y is used, then the correct value of 104 results for the atomic number of Rutherfordium. However, if instead the Einstein definition of Y is used, then the predicted atomic number of Rutherfordium is wrongly calculated to be 107. As I say, it's no surprise that Naughton-Lorentz predicts the correct result since it is derived from this same calculation, but this does flag up a possible slight discrepancy in Einstein's definition.

3. A Possible Explanation Of The Einstein Discrepancy

The Pythagoras theorem was used to derive Einstein relativity. In doing so it was assumed that the dimensions of space are _always perpendicular to each other._

In appendix 2 (see below) the more general cosine rule was used to derive an angle D in degrees between the dimensions of space which would yield values for the Lorentz factor Y_L which would agree with the ionization energies of the last electron in each atom.

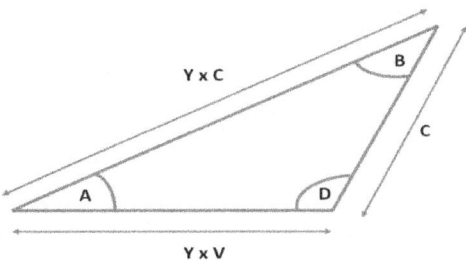

Diagram 3.1 The Relativity Triangle

This diagram shows a beam of light travelling at speed 'c' opposite angle A as observed by a passenger on a train and the relative speed of that train as observed by a stationary observer converted to the same timeframe as the passenger Y x V opposite angle B and similarly the beam of light as observed by the stationary person but converted to the timeframe of the passenger as Y x C opposite angle D.

It was found by plotting a graph of angle D v V/c;-

$$D = 116.81V^6/c^6 - 185.33V^5/c^5 + 115.9V^4/c^4 - 33.327V^3/c^3$$
$$+ 4.2384V^2/c^2 - 0.2044V/c + 90.00 \qquad ...(29)$$

If correct, this formula suggests that the angle D would be 108.1° when the velocity reaches the speed of light c.

It can be seen from graph 2.2 that the angle D is 90° when the velocity V is zero. As the velocity is increased the angle D decreases, reaching a minimum of D = 89.995 when V = 0.2444c

After this velocity the angle D increases again until at the point when V = 0.3531c (at this point the last electron in the atom is travelling at $1.0587x10^8$ m/s when the number of protons in the tin atom is P=50) the angle D reaches 90° again.

15

Above this velocity the angle D gradually increases above 90° until it reaches 108.1° when the velocity V reaches the speed of light i.e. V=C

This might help to explain how nature appears to have a built-in directional bias. For example, the left-hand motor rule indicates that when a current flows in a wire that is perpendicular to a magnetic field then the wire experiences a force in a particular direction. *But what is it that causes the centripetal force to experience a particular distortion depending on the velocity of the electron?*

The angle D initially decreases below 90° as the velocity of the single electron increases but eventually the angle increases again and exceeds 90° This can be seen in the graph below which shows how the angle D (which Einstein assumed was always 90°) varies with the number of protons P in an atom with only one remaining electron. In this graph only the values of angle D for atoms with 65 protons or less are shown so that the values of D below 90° an be clearly seen for the lower values of P.

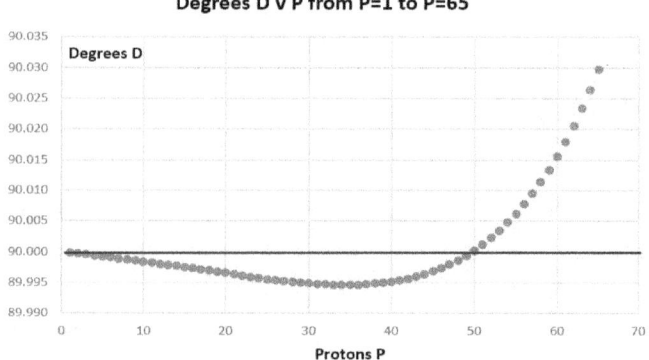

Graph 3 - Angle D v P protons for the last electron in each atom.

Einstein described the distortion of spacetime to account for the effects of gravity. The results presented here suggest that spacetime *might* appear distorted to an object moving at velocity V. Of course there might be other factors at play here.

Angle D reaches a minimum D=89.995° when P=34 (Selenium). The angle increases again after this point until it reaches 90° again when P=50 (Tin). As P is increased further above this point, the angle D exceeds 90° until is reaches a maximum of 108.1° when the velocity of the last remaining electron reaches c the speed of light when P is at a value of P =170

The fact that the angle decreases at first and then increases again at higher values of velocity might suggest there is more than one contributing factor. Could the variation in the size of angle D be caused by the distortion of the fabric of spacetime? Alternatively, could the spin of the electron vary as its speed increases? It has been assumed here that electrons exist as solid 'dots' that travel in circular orbits around atoms. Of course this is a massive over simplification. It's generally accepted that electrons exist as clouds of energy which are more spread out and that their orbits are quite likely to be elliptical. Quite possibly the geometry of these orbits might change as the speed at which an electron travels is increased.

Perhaps the increasing number of protons in the nucleus of the atom around which the last remaining electron travels, might cause some kind of additional interaction or lead to other impacts perhaps caused by effects such as electromagnetism.

It's not in any way obvious what effects might cause the variation in the angle D with velocity V. It may be something which is only seen in the case of single electrons orbiting around atoms or it might actually be something which is observed more widely. Further investigation would be needed to properly understand this effect. It would be interesting to study fast moving objects on a different scale, perhaps quasars, to try to identify if any of these effects could result from the impacts of relativity.

Although no definite conclusion is reached in this work regarding the precise cause of the apparent distortion in spacetime described above, it might be possible to shed some light on this matter by looking at what would happen to the Lorentz coefficient Y if the speed of the electron V exceeded the speed of light c. The results of this calculation are shown below;-

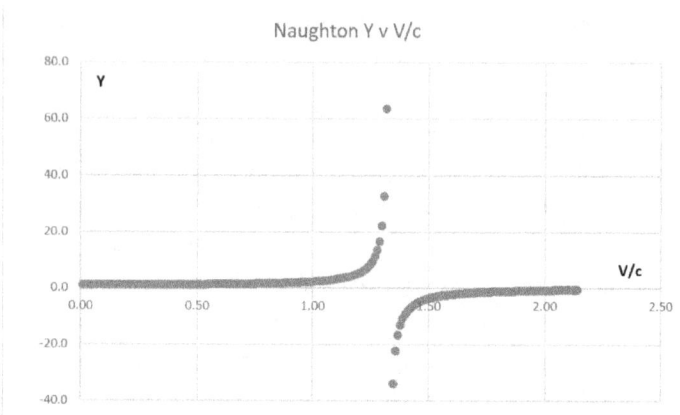

Graph 4 - The Calculated Naughton-Lorentz Coefficient Y v Velocity v/c for velocity $V <= 2^{0.5}$

At velocities below the speed of light (V < c) the value of the Lorentz coefficient Y increases from 1 when V = 0 up to Y = 2.298 when V = c (*N.B. the results derived in this study of the energies of the last electron in each atom differ significantly from the equivalent Einstein calculation*)

There is a *discontinuity* at the point where the velocity reaches 1.33 times the speed of light (V = 1.33c) but beyond this point the value of Y turns negative. e.g. at velocity V = 1.32c the angle D is predicted to be 247° and the Lorentz coefficient Y becomes 63.62
As the velocity increases further the magnitude of the Y coefficient drops to Y = -69.76 when V = 1.34c at which point the angle D = 266°

As V is increased further beyond V = 1.34c the value of Y decreases further until at the point V = $2^{1/2}$ x c (i.e. V = 1.414c) the angle D = 353.6° (almost a complete circle) and the value of Y = -7.7

Equation 16 suggests that the universal speed limit (the maximum velocity achievable) might be significantly greater than the speed of light c proposed by Einstein.

Instead, it suggests the universal speed limit might be V = 1.33 x c

This might better explain the observed expansion of the universe, over the period of its development since the occurrence of the Big Bang. If the deceleration of the universe has remained constant (as suggested by my interpretation of Hubble's law - see An Alternative History Of Time) then the expansion speed of the outer edge of the universe would have gradually slowed down as it aged, but would not have slowed down to the speed of light c until the universe was about 4.6 billion years old assuming it's initial speed was the universal speed limit of 1.33c as proposed here.

It's generally accepted that the Dark Energy phase of the universe occurred at a time about 9.8 billion years after the occurrence of the Big Bang, at which point the expansion speed of the outer edge of the universe would have slowed to about V = 0.63c (according to the deceleration suggested by Hubble's constant) *at which point angle D would have reduced to about 90° for the first time.*

The current proposed explanations for the onset of the Dark Matter phase tend to rely on the evolution of the separate four major forces, but there doesn't appear to be any compelling proposal to explain why this separation actually occurred. *Might it be the case that the onset of the Dark Matter phase occurred at the exact point when the speed of expansion of the edge of the universe slowed down sufficiently for the angle D to reduce to 90°?*

This mechanism might also go someway to explain the lack of the existence of observable antimatter in the universe.

It's interesting to calculate what happens to the other angles in the relativity triangle at different values of velocity V.

The sine rule can be used to calculate angle B.

Since;-

$$sin(B)/(Y \times V) = sin(D)/(Y \times c) \qquad ...(30)$$

$$sin(B) = (V/C) \times sin(D) \qquad ...(31)$$

And having calculated angles B and D we can *estimate* angle A since the angles in a two-dimensional triangle add up to 180°

The results plotted against V/C are as follows;-

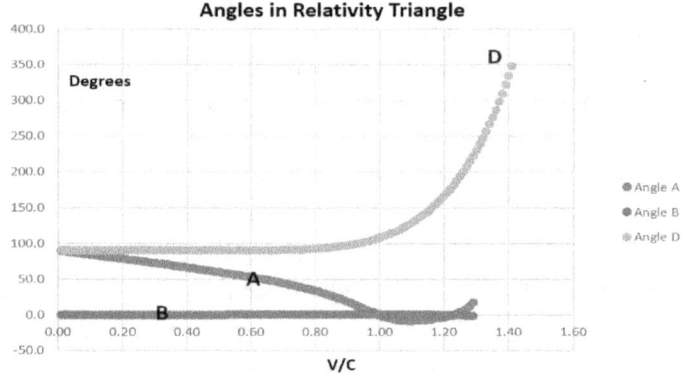

Graph 5 - The Angles in the Relativity Triangle plotted against V/C

Angle A starts at 90° when V/C = 0
It decreases until it reaches 0° when V/C = 1
When V/C = 1.09 it reaches a minimum of -7.86° and then increases back to 0° when V/C = 1.22

It finally reaches a maximum value of 18° when V/C = 1.29

Conversely angle B starts at 0° when V/C = 0
It reaches a maximum 72° when V/C = 1.01 and then reduces to 0° when V/C = 1.22

Beyond this angle B reduces further until it reaches -60° when V/C = 1.29

Beyond this, the magnitude of sin(B) exceeds the value 1 [sin(B) = -1.002 when V/C = 1.30]

If it is really the case that immediately after the Big Bang the maximum speed at which the initial boundary of the universe moved outwards was 1.33c, then according to the constant Hubble deceleration (assuming a = $-k_Hc$ where k_H is Hubble's constant - *see An Alternative History of Time*) then the universe should stop expanding after about 18.5 billion years having reached a maximum radius of 12.3 billion light years.

If the initial speed at which the outer boundary of the universe moved immediately after the Big Bang was c then by now (after 13.8 billion years of constant Hubble deceleration) its speed would have reduced to 1% of c

However, if the initial speed of the expansion of the universe was 1.33c then the outer boundary should still be moving at about 34% of c. It would be challenging indeed to measure the actual speed of the boundary of the universe but this could be helpful in helping us to fully understand the changes likely to be experienced as the universe ages further.

4. Conclusions

An alternative definition of the Lorentz coefficient has been derived from the ionization energies of the last electron in each atom.

$$Y_{NAUGHTON} = 1/(1 - 0.5649 \times V^2/C^2)$$

The *apparent* angle D in degrees between the dimensions of space are relative to the velocity V of a moving object and is given by;-

$$D = 116.81V^6/C^6 - 185.33V^5/C^5 + 115.9V^4/C^4 - 33.327V^3/C^3$$
$$+ 4.2384V^2/C^2 - 0.2044V/C + 90.00$$

At the time of writing these proposals are unconfirmed. It remains to be seen whether these proposals will be proven or not.

Appendices

Appendix 1 - Einstein Calculated The Relativity Of Time

Albert Einstein fundamentally changed our understanding of time when he derived the relationship between the *relative* rate at which time passes and the *relative* velocity of any particular object.

Einstein derived an expression for time 't' by applying the results of the Michelson–Morley experiment. In 1887 Albert A. Michelson and Edward W. Morley published results of an experiment which measured the speed of light in perpendicular directions in an attempt to measure the speed of Earth through the aether, an imagined substance which it was thought must exist to provide a medium through which light waves could propagate. Before Einstein it was thought that light was a wave and as such it must travel in a medium, just like a water wave can only travel in water then it was thought that light must also need a material to travel through and this imaginary material was termed the aether. No one had ever detected the presence of the aether but by comparing the speed of light at right angles, Michelson and Morley were expecting to measure the speed at which the aether was flowing relative to the motion of the Earth travelling around the sun. They carried out precise measurements over a period of about ten years.

The conclusion from Michelson and Morley was that the speed of light 'c' was constant in all directions for all observers irrespective of their velocity. This result was a true watershed in the understanding of the propagation of light. It provided compelling proof that the aether did not exist. In 1905 Einstein took the idea of the constant speed of light and applied it to derive an understanding of the way in which the measurement of time must vary as experienced by objects moving at differing speeds relative to each other.

Einstein considered how the constant speed of light would affect observations of the same event by people in different locations. First he considered a fast-moving train, moving at constant high speed in a straight line. He imagined a passenger seated on the train looking at a beam of light moving towards a window on the train from inside the train.

Einstein analyzed what the passenger on the train would observe (we could call the passenger observer1). Einstein compared what the passenger on the train might see compared to what an observer at the side of the railway track might witness, bearing in mind that the speed of light must remain constant for both observers in order to comply with the conclusions which Michelson and Morley had derived.

First let's consider what the observer on the train would see. Diagram A1.1 below shows the passenger on the train seeing a beam of light travelling from his position 'a' to the window at position 'b' which is a distance h_1

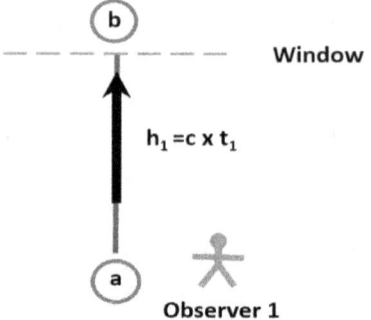

Diagram A1.1 - A Passenger On A Train Looking At A Beam Of Light
The time taken for a photon of light to travel distance h_1 is given by time t_1 so we can say;-

$$h_1 = c \times t_1 \qquad \text{... A1.1}$$

This seems a fairly uncontroversial assertion given that we define velocity as a ratio of distance travelled in a certain time t and we know that light must travel at constant speed 'c'.

So next Einstein considered what an observer at the side of the track would see. The observations of this second observer are shown in diagram A1.2 below;-

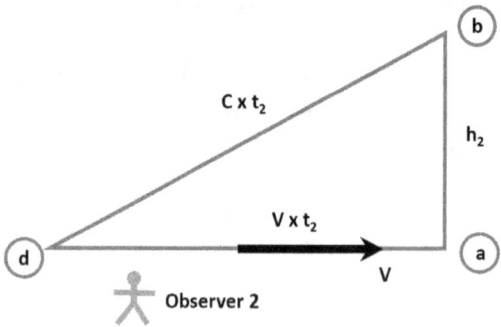

Diagram A1.2 - Observer 2 At The Side Of The Track Sees A Fast Moving Train Travel Past At Velocity V

General Relativity Revisited

The case described in diagram A1.2 is exactly the same case as was described in diagram A1.1, we are talking about the same train moving at the same velocity V, except this time the observer is standing at the side of the track and sees the train travel from point 'd' to point 'a' along the track. This time the observer sees the train travel from position 'd' to position 'a' in time t_2. The distance d to a must be;-

$$L_{da} = V \times t_2 \qquad \text{... A1.2}$$

This observer at the side of the track will see the same light beam that the passenger on the train sees but this time the light beam will appear to travel from position d to position b in time t_2. Of course, the speed of light is still the universal constant 'c' so we can say;-

$$L_{db} = c \times t_2 \qquad \text{... A1.3}$$

Applying Pythagoras's theorem we can say;-

$$(L_{db})^2 = (L_{da})^2 + (L_{ab})^2 \quad \text{... A1.4}$$

Therefore;-

$$(c \times t_2)^2 = (V \times t_2)^2 + (h_2)^2 \quad \text{...A1.5}$$

According to Einstein distance h_1 and distance h_2 should be identical since the entire movement of the train has been at right angles to this plane so using this relationship we can say;-

$$h_1 = h_2 \quad \text{... A1.6}$$

Substituting from equations A1.1 and A1.6 into equation A1.5 we can therefore say;-

$$(c \times t_1)^2 = (c^2 - V^2) \times t_2^2 \quad \text{... A1.7}$$

Rearranging this we can say;-

$$t_2 = c \times t_1 / (c^2 - V^2)^{0.5} \quad ... \text{A1.8}$$

This can be expressed as;-

$$t_2 = Y \times t_1 \quad\quad\quad ... \text{A1.9}$$

Where Y is known as the Lorentz coefficient expressed as;-

$$Y = c / (c^2 - V^2)^{0.5} \quad ... \text{A1.10}$$

Einstein deduced that the Kinetic Energy of any object moving at velocity V can be expressed as;

$$E = mc^2(Y - 1) \quad\quad ... \text{A1.11}$$

where m is the rest mass of the particular object in question. All of this has been tested by many hundreds of experiments of various kinds and these equations describe to a remarkable degree of accuracy the observed behavior of most objects, even ones moving at very high speeds. For example, Einstein's calculations were the very first to adequately explain the time it took for planet Mercury to complete its rotational orbit around the sun. Mercury takes 87.969 days to travel around the sun which is about 20 minutes (0.014 days) longer than the time predicted by classical Physics.

Appendix 2 - The Distortion Of Space-Time

Einstein used the Pythagoras theorem to derive his theory of General Relativity. In doing so he assumed that space was perfectly square at all times - that the dimensions of space were perfectly perpendicular to each other. However, the analysis of the ionization energy of the last electron in each atom performed here in this work, suggests that there might be a discrepancy in Einstein's theory. It might be possible to explain this variance if it's assumed the dimensions of space are not necessarily perpendicular.

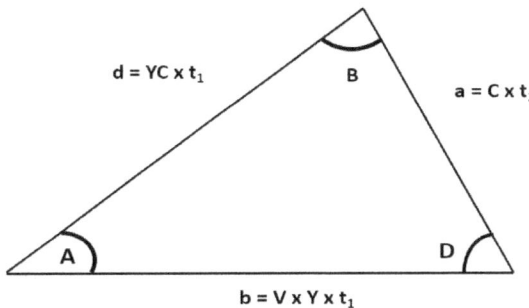

Diagram A2.1 The Cosine Rule

The cosine rule provides a more general relationship between the lengths of the sides of a triangle.

$$d^2 = a^2 + b^2 - 2abx\cos(D) \quad ... A2.1$$

where D is the angle opposite side d in Diagram A2.1

Of course when angle $D = 90°$ then $\cos(D) = 0$
In these circumstances the cosine rule is consistent with the Pythagoras theorem.

If we apply the definitions of each dimension used by Einstein to the cosine rule we get;-

$$b = V \times t_2 \quad ... A2.2$$

where 'b' is the distance travelled the fast-moving train in time t_2 speeding along the track at velocity V as observed by the person standing at the side of the track

$$a = c \times t_1 \quad ... A2.3$$

where 'a' is the distance travelled by a beam of light in time t_1 as observed by a passenger on the train.

$$d = c \times t_2 \quad ... A2.4$$

where d is the distance travelled by the beam of light in time t_2 as observed by the person standing at the side of the track.

NOTE: The angle opposite side d is angle D which is not necessarily 90° *although it was assumed by Einstein that this angle would always be 90°*

Substituting equations A2.2, A2.3 and A2.4 into A2.1 we can say;-

$$c^2t_2^2 = v^2t_2^2 + c^2t_1^2 - 2cvt_1t_2\cos(D) \quad\quad ... A2.5$$

Since $t_2 = Y \times t_1$ we can say;-

$$\cos(D) = \{c^2 + Y^2(V^2 - c^2)\}/2cVY \quad\quad ... A2.6$$

Hence the value of angle D was calculated for each of the atoms containing one remaining electron and P protons. A graph of D was plotted against the number of protons P in each atom.

The results are as follows;-

General Relativity Revisited

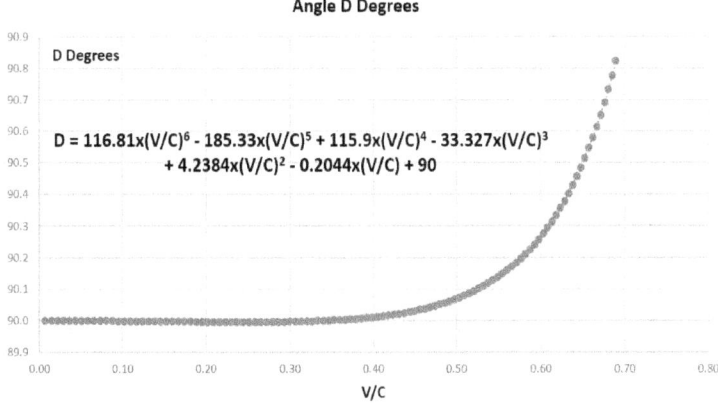

Graph A2.1 Naughton Angle D versus V/C where V is the velocity of the last electron in each atom.

This curve dips below 90° for atoms with less than 50 protons. This can be seen more clearly if we plot angle D against the number of protons P just for elements with 65 protons or less.

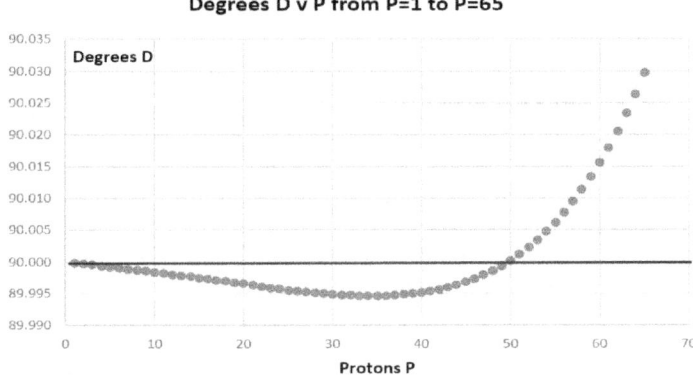

Graph A2.2 Naughton Angle D versus P protons for the last electron in elements with 65 protons or less.

Element – Measured First Ionizations

P	Element	Ion Energy /J EION	Ion Energy /eV	D /Degrees
1	Hydrogen	2.1786616E-18	13.598	89.9998
2	Helium	8.7199129E-18	54.425	89.9997
3	Lithium	1.9622088E-17	122.470	89.9995
4	Berylium	3.4887291E-17	217.747	89.9994
5	Boron	5.4517849E-17	340.270	89.9992
6	Carbon	7.8516584E-17	490.057	89.9990
7	Nitrogen	1.0688765E-16	667.134	89.9989
8	Oxygen	1.3963486E-16	871.524	89.9987
9	Fluorine	1.7676370E-16	1103.261	89.9985
10	Neon	2.1827933E-16	1362.379	89.9984
11	Sodium	2.6418820E-16	1648.917	89.9982
12	Magnesium	3.1449764E-16	1962.920	89.9980
13	Aluminium	3.6921578E-16	2304.441	89.9978
14	Silicon	4.2835109E-16	2673.530	89.9977
15	Phosphorous	4.9191237E-16	3070.245	89.9975
16	Sulphur	5.5990959E-16	3494.646	89.9973
17	Chlorine	6.3235437E-16	3946.806	89.9971
18	Argon	7.0926000E-16	4426.808	89.9969
19	Potassium	7.9063146E-16	4934.684	89.9968
20	Calcium	8.7649200E-16	5470.578	89.9966
21	Scandium	9.6684329E-16	6034.501	89.9964
22	Titanium	1.0615840E-15	6625.820	89.9960
23	Vanadium	1.1609681E-15	7246.120	89.9959
24	Chromium	1.2649007E-15	7894.810	89.9957

25	Manganese	1.3733900E-15	8571.940	89.9955
26	Iron	1.4864647E-15	9277.690	89.9954
27	Cobalt	1.6041346E-15	10012.120	89.9952
28	Nickel	1.7264267E-15	10775.400	89.9951
29	Copper	1.8533552E-15	11567.617	89.9950
30	Zinc	1.9849450E-15	12388.928	89.9948
31	Gallium	2.1212211E-15	13239.488	89.9948
32	Germanium	2.2622046E-15	14119.429	89.9947
33	Arsenic	2.4079203E-15	15028.906	89.9946
34	Selenium	2.5583946E-15	15968.083	89.9946
35	Bromine	2.7136540E-15	16937.126	89.9946
36	Krypton	2.8737262E-15	17936.208	89.9946
37	Rubidium	3.0386410E-15	18965.515	89.9947
38	Strontium	3.2084282E-15	20025.233	89.9948
39	Yttrium	3.3831180E-15	21115.550	89.9949
40	Zirconium	3.5627442E-15	22236.677	89.9951
41	Niobium	3.7473365E-15	23388.800	89.9953
42	Molybdenum	3.9369326E-15	24572.154	89.9956
43	Technetium	4.1315726E-15	25786.989	89.9959
44	Ruthenium	4.3312878E-15	27033.501	89.9963
45	Rhodium	4.5361222E-15	28311.964	89.9968
46	Palladium	4.7461113E-15	29622.600	89.9973
47	Silver	4.9613014E-15	30965.697	89.9979
48	Cadmium	5.1817299E-15	32341.490	89.9986
49	Indium	5.4074500E-15	33750.310	89.9994
50	Tin	5.6384990E-15	35192.390	90.0002
51	Antimony	5.8749282E-15	36668.050	90.0012
52	Tellurium	6.1167808E-15	38177.560	90.0023
53	Iodine	6.3641353E-15	39721.410	90.0034
54	Xenon	6.6170094E-15	41299.710	90.0047
55	Caesium	6.8754879E-15	42912.990	90.0062

56	Barium	7.1396062E-15	44561.470	90.0077
57	Lanthanum	7.4094363E-15	46245.600	90.0095
58	Cerium	7.6850326E-15	47965.720	90.0113
59	Praesodymium	7.9664626E-15	49722.250	90.0134
60	Neodymium	8.2537886E-15	51515.580	90.0156
61	Promethium	8.5470732E-15	53346.100	90.0180
62	Samarium	8.8463836E-15	55214.230	90.0206
63	Europium	9.1518256E-15	57120.630	90.0234
64	Gadolonium	9.4634345E-15	59065.520	90.0264
65	Terbium	9.7813288E-15	61049.640	90.0297
66	Dysprosium	1.0105590E-14	63073.500	90.0332
67	Holmium	1.0436171E-14	65136.800	90.0369
68	Erbium	1.0773432E-14	67241.800	90.0410
69	Thulium	1.1117183E-14	69387.300	90.0454
70	Ytterbium	1.1467662E-14	71574.800	90.0500
71	Lutetium	1.1824951E-14	73804.800	90.0550
72	Hafnium	1.2189130E-14	76077.800	90.0603
73	Tantalum	1.2560342E-14	78394.700	90.0660
74	Tungsten	1.2938603E-14	80755.600	90.0720
75	Rhenium	1.3324203E-14	83162.300	90.0785
76	Osmium	1.3717077E-14	85614.400	90.0854
77	Iridium	1.4117449E-14	88113.300	90.0928
78	Platinum	1.4525431E-14	90659.700	90.1006
79	Gold	1.4941136E-14	93254.300	90.1089
80	Mercury	1.5364659E-14	95897.700	90.1177
81	Thallium	1.5796274E-14	98591.600	90.1270
82	Lead	1.6236044E-14	101336.400	90.1370
83	Bismuth	1.6684081E-14	104132.800	90.1475
84	Polonium	1.7140690E-14	106982.700	90.1587
85	Astatine	1.7605855E-14	109886.000	90.1705
86	Radon	1.8079735E-14	112843.700	90.1830

General Relativity Revisited

87	Francium	1.8562844E-14	115859.000	90.1963
88	Radium	1.9055086E-14	118931.300	90.2104
89	Actinium	1.9556829E-14	122062.900	90.2252
90	Thorium	2.0068008E-14	125253.400	90.2409
91	Protoactinium	2.0589314E-14	128507.100	90.2576
92	Uranium	2.1120296E-14	131821.200	90.2751
93	Neptunium	2.1662030E-14	135202.400	90.2937
94	Plutonium	2.2213761E-14	138646.000	90.3132
95	Americium	2.2776932E-14	142161.000	90.3340
96	Curium	2.3350837E-14	145743.000	90.3558
97	Berkelium	2.3936438E-14	149398.000	90.3790
98	Californium	2.4533415E-14	153124.000	90.4033
99	Einsteinium	2.5142569E-14	156926.000	90.4289
100	Fermium	2.5763899E-14	160804.000	90.4560
101	Mendelevium	2.6398368E-14	164764.000	90.4847
102	Nobelium	2.7045974E-14	168806.000	90.5149
103	Lawrencium	2.7706718E-14	172930.000	90.5467
104	Rutherfordium	2.8382523E-14	177148.000	90.5806
105	Dubnium	2.9070985E-14	181445.000	90.6158
106	Seaborgium	2.9774348E-14	185835.000	90.6529
107	Bohrium	3.0494373E-14	190329.000	90.6925
108	Hassium	3.1228498E-14	194911.000	90.7338
109	Meitnerium	3.1980567E-14	199605.000	90.7778
110	Darmstadtium	3.2747857E-14	204394.000	90.8237

Patrick Naughton

Appendix 3 - Ionisation Energies Last Electrons

The ionisation energies E_{ion} for the *last* electrons in each atom are listed in the table below.

P is the number of Protons in the atom.

Y is the Lorentz coefficient calculated from the relationship

$E = mc^2(Y-1)$

Measured Energy

| | | Ion Energy | Ion Energy | |
| | | Measured | Measured | |
P	Element	/eV	/Joules	Y
1	Hydrogen	13.60	2.179E-18	1.00003
2	Helium	54.42	8.720E-18	1.00011
3	Lithium	122.47	1.962E-17	1.00024
4	Beryllium	217.75	3.489E-17	1.00043
5	Boron	340.27	5.452E-17	1.00067
6	Carbon	490.06	7.852E-17	1.00096
7	Nitrogen	667.13	1.069E-16	1.00131
8	Oxygen	871.52	1.396E-16	1.00171
9	Fluorine	1,103.26	1.768E-16	1.00216
10	Neon	1,362.38	2.183E-16	1.00267
11	Sodium	1,648.92	2.642E-16	1.00323
12	Magnesium	1,962.92	3.145E-16	1.00384
13	Aluminium	2,304.44	3.692E-16	1.00451
14	Silicon	2,673.53	4.284E-16	1.00523

General Relativity Revisited

P	Element	/eV	/Joules	Y
15	Phosphorous	3,070.24	4.919E-16	1.00601
16	Sulphur	3,494.65	5.599E-16	1.00684
17	Chlorine	3,946.81	6.324E-16	1.00772
18	Argon	4,426.81	7.093E-16	1.00866
19	Potassium	4,934.68	7.906E-16	1.00966
20	Calcium	5,470.58	8.765E-16	1.01071
21	Scandium	6,034.50	9.668E-16	1.01181
22	Titanium	6,625.82	1.062E-15	1.01297
23	Vanadium	7,246.12	1.161E-15	1.01418
24	Chromium	7,894.81	1.265E-15	1.01545
25	Manganese	8,571.94	1.373E-15	1.01678
26	Iron	9,277.69	1.486E-15	1.01816
27	Cobalt	10,012.12	1.604E-15	1.01959
28	Nickel	10,775.40	1.726E-15	1.02109
29	Copper	11,567.62	1.853E-15	1.02264
30	Zinc	12,388.93	1.985E-15	1.02424
31	Gallium	13,239.49	2.121E-15	1.02591
32	Germanium	14,119.43	2.262E-15	1.02763
33	Arsenic	15,028.91	2.408E-15	1.02941
34	Selenium	15,968.08	2.558E-15	1.03125
35	Bromine	16,937.13	2.714E-15	1.03315
36	Krypton	17,936.21	2.874E-15	1.03510
37	Rubidium	18,965.52	3.039E-15	1.03711
38	Strontium	20,025.23	3.208E-15	1.03919

P	Element	/eV	/Joules	Y
39	Yttrium	21,115.55	3.383E-15	1.04132
40	Zirconium	22,236.68	3.563E-15	1.04352
41	Niobium	23,388.80	3.747E-15	1.04577
42	Molybdenum	24,572.15	3.937E-15	1.04809
43	Technetium	25,786.99	4.132E-15	1.05046
44	Ruthenium	27,033.50	4.331E-15	1.05290
45	Rhodium	28,311.96	4.536E-15	1.05541
46	Palladium	29,622.60	4.746E-15	1.05797
47	Silver	30,965.70	4.961E-15	1.06060
48	Cadmium	32,341.49	5.182E-15	1.06329
49	Indium	33,750.31	5.407E-15	1.06605
50	Tin	35,192.39	5.638E-15	1.06887
51	Antimony	36,668.05	5.875E-15	1.07176
52	Tellurium	38,177.56	6.117E-15	1.07471
53	Iodine	39,721.41	6.364E-15	1.07773
54	Xenon	41,299.71	6.617E-15	1.08082
55	Caesium	42,912.99	6.875E-15	1.08398
56	Barium	44,561.47	7.140E-15	1.08721
57	Lanthanum	46,245.60	7.409E-15	1.09050
58	Cerium	47,965.72	7.685E-15	1.09387
59	Praesodymium	49,722.25	7.966E-15	1.09730
60	Neodymium	51,515.58	8.254E-15	1.10081
61	Promethium	53,346.10	8.547E-15	1.10440
62	Samarium	55,214.23	8.846E-15	1.10805
63	Europium	57,120.63	9.152E-15	1.11178
64	Gadolinium	59,065.52	9.463E-15	1.11559

General Relativity Revisited

P	Element	/eV	/Joules	Y
65	Terbium	61,049.64	9.781E-15	1.11947
66	Dysprosium	63,073.50	1.011E-14	1.12343
67	Holmium	65,136.80	1.044E-14	1.12747
68	Erbium	67,241.80	1.077E-14	1.13159
69	Thulium	69,387.30	1.112E-14	1.13579
70	Ytterbium	71,574.80	1.147E-14	1.14007
71	Lutetium	73,804.80	1.182E-14	1.14443
72	Hafnium	76,077.80	1.219E-14	1.14888
73	Tantalum	78,394.70	1.256E-14	1.15342
74	Tungsten	80,755.60	1.294E-14	1.15804
75	Rhenium	83,162.30	1.332E-14	1.16275
76	Osmium	85,614.40	1.372E-14	1.16754
77	Iridium	88,113.30	1.412E-14	1.17244
78	Platinum	90,659.70	1.453E-14	1.17742
79	Gold	93,254.30	1.494E-14	1.18250
80	Mercury	95,897.70	1.536E-14	1.18767
81	Thallium	98,591.60	1.580E-14	1.19294
82	Lead	101,336.40	1.624E-14	1.19831
83	Bismuth	104,132.80	1.668E-14	1.20378
84	Polonium	106,982.70	1.714E-14	1.20936
85	Astatine	109,886.00	1.761E-14	1.21504
86	Radon	112,843.70	1.808E-14	1.22083
87	Francium	115,859.00	1.856E-14	1.22673
88	Radium	118,931.30	1.906E-14	1.23275
89	Actinium	122,062.90	1.956E-14	1.23887
90	Thorium	125,253.40	2.007E-14	1.24512

Patrick Naughton

P	Element	/eV	/Joules	Y
91	Protoactinium	128,507.10	2.059E-14	1.25148
92	Uranium	131,821.20	2.112E-14	1.25797
93	Neptunium	135,202.40	2.166E-14	1.26459
94	Plutonium	138,646.00	2.221E-14	1.27133
95	Americium	142,161.00	2.278E-14	1.27820
96	Curium	145,743.00	2.335E-14	1.28521
97	Berkelium	149,398.00	2.394E-14	1.29237
98	Californium	153,124.00	2.453E-14	1.29966
99	Einsteinium	156,926.00	2.514E-14	1.30710
100	Fermium	160,804.00	2.576E-14	1.31469
101	Mendelevium	164,764.00	2.640E-14	1.32244
102	Nobelium	168,806.00	2.705E-14	1.33035
103	Lawrencium	172,930.00	2.771E-14	1.33842
104	Rutherfordium	177,148.00	2.838E-14	1.34667

References

1. "LHC sets new world record". Media and Press Relations (Press release). CERN. 30 November 2009. Retrieved 13 November 2016.

/ends

Patrick Naughton